SPOTLIGHT ON GLOBAL ISSUES

ENERGY SOLUTIONS FOR ALL

Adam Furgang

ROSEN PUBLISHING
NEW YORK

Published in 2022 by The Rosen Publishing Group, Inc.
29 East 21st Street, New York, NY 10010

First Edition

Editor: Theresa Emminizer
Book Design: Michael Flynn

Photo Credits: Cover Jenson/Shutterstock.com; (series globe background) photastic/Shutterstock.com; p. 4 solarseven/Shutterstock.com; p. 5 Buyenlarge/Archive Photos/Getty Images; p. 6 lusia83/Shutterstock.com; p. 7 Bettmann/Getty Images; p. 8 Vijay Talla/EyeEm/Getty Images; p. 9 Funny Solution Studio/Shutterstock.com; p. 11 U.S. Coast Guard/Getty Images; p. 12 Sunshine Seeds/Shutterstock.com; p. 13 Procyk Radek/Shutterstock.com; p. 14 Tatiana Grozetskaya/Shutterstock.com; p. 15 https://commons.wikimedia.org/wiki/File:William_Ruckelshaus_Swearing_In_as_EPA_Administrator.jpg; p. 16 https://commons.wikimedia.org/wiki/File:Donald_Trump_official_portrait.jpg; p. 17 Pacific Press/LightRocket/Getty Images; p. 18 Hulton Archive/Getty Images; p. 19 Library of Congress/Corbis Historical/Getty Images; p. 21 Bloomberg/Getty Images; p. 22 taraki/Shutterstock.com; p. 23 NASA/Getty Images; p. 25 Ververidis Vasilis/Shutterstock.com; p. 26 Noam Galai/Getty Images; p. 27 Holli/Shutterstock.com; p. 28 Jacek Chabraszewski/Shutterstock.com; p. 29 BlurryMe/Shutterstock.com.

Cataloging-in-Publication Data

Names: Furgang, Adam.
Title: Energy solutions for all / Adam Furgang.
Description: New York : Rosen Publishing, 2022. | Series: Spotlight on global issues | Includes glossary and index.
Identifiers: ISBN 9781725323292 (pbk.) | ISBN 9781725323322 (library bound) | ISBN 9781725323308 (6 pack)
Subjects: LCSH: Power resources--Juvenile literature. | Renewable energy sources--Juvenile literature. | Energy consumption--Juvenile literature.
Classification: LCC TJ163.23 F87 2022 | DDC 333.79--dc23

Manufactured in the United States of America

Some of the images in this book illustrate individuals who are models. The depictions do not imply actual situations or events.

CPSIA Compliance Information: Batch #CSR22. For further information contact Rosen Publishing, New York, New York at 1-800-237-9932.

CONTENTS

CHAPTER ONE

A BRIEF HISTORY OF ENERGY

Hundreds of thousands of years ago, people used wood as one of the first energy sources. They burned it to create fire for warming themselves, cooking food, and protecting themselves from dangerous animals. Since then, people have used wood and other energy sources for these and many other purposes. As societies grew and energy demands increased, they needed greater sources of energy.

Fossil fuels formed deep underground over millions of years. Since the start of the Industrial Revolution in the late 1700s, people have used fossil fuels such as coal, natural gas, and oil to power new forms of technology, such as the steam engine and, later, the electric turbine and other machines.

These wind turbines in Missouri generate electrical power. Wind is a naturally occurring renewable source of energy. It can be continually used without running out.

Despite the many benefits of using fossil fuels, there are drawbacks as well. Burning fossil fuels to generate electricity and run vehicles releases large amounts of carbon dioxide (CO_2) and other gases that trap heat in Earth's atmosphere. This has increased the overall temperature on Earth, which is harmful to the environment. Growing climate changes and limited fossil fuel resources have led us to look for new, renewable energy sources that are less harmful to the planet.

Around the world, renewable kinds of energy such as wind, solar, geothermal, hydroelectric, and nuclear energy are increasingly used in place of fossil fuel energy. However, in 2018, world oil consumption was 96.9 million barrels a day. By 2040, it's estimated that that number will rise to 106.4 million barrels a day. To curb the effects of burning fossil fuels, seeking out clean energy sources is more important than ever.

Energy sources have changed over time. Prior to the 1700s and the start of the Industrial Revolution, common energy sources included wood for heat and cooking and animals for labor. People used wind power and waterpower for tasks such as pumping water and grinding grain. Wind was also used to propel boats and ships with sails.

Once the Industrial Revolution was underway in the 1800s, coal use increased rapidly. Coal is heavier than wood and burns longer, making it a useful fuel to power machines such as steam engines and electric generators. Coal is a sedimentary rock that was created millions of years ago through the natural process of pressure and heat on organic plant matter. It's a nonrenewable fossil fuel. Today, coal generates 38 percent of the electricity in the world.

Petroleum, or crude oil, was the next energy source used for fuel. Oil, like coal, is a nonrenewable fossil fuel that comes from underground. People can use oil to make other fuels, such as gasoline for cars.

People knew about oil before it became widely used. In 1859, American businessman Edwin Drake was the first to drill a successful oil well. In 1901, a drilling location called Spindletop Hill in southeastern Texas erupted with oil. The oil shot 150 feet (45.7 m) in the air and the well produced almost 100,000 barrels of oil a day. Since then, worldwide oil use has grown at an incredible pace to create fuel and other materials, such as plastic.

COAL

Edwin Drake (right) is pictured here standing in front of his oil well.

CHAPTER TWO

RENEWABLE ENERGY SOURCES

Modern societies depend on electricity to power everything from washing machines to cell phones. As the world population has increased, so has global energy use. Since 1800, the world population has grown from around 1 billion people to over 7 billion people. In 1859, the United States produced 2,000 barrels of oil in one year. In April 2019, U.S. oil production hit a record of 12 million barrels per day. However, because using fossil fuels contributes to climate change, people are increasingly trying to use clean, renewable sources of energy.

Hydropower is a renewable way of generating electricity that produces few greenhouse gas emissions. Naturally moving water from rivers or dams is channeled to spin turbines that generate electricity. In 2015, hydropower generated 16.6 percent of the world's electricity.

Modern cities around the world, such as Dubai, United Arab Emirates, need electricity to run continually. Fossil fuels, such as oil and coal, are the primary fuel sources for electricity worldwide.

A sustainable energy source may not be renewable forever, but it can last a very long time. One sustainable kind of energy is atomic energy, or energy released by nuclear fission, splitting atoms to release large amounts of heat energy. Nuclear reactors use these controlled atomic reactions to generate heat, which creates steam from water, which powers generators that produce electricity. In 1942, physicist Enrico Fermi helped build the world's first nuclear reactor, the Chicago Pile 1 Reactor. Today, nuclear sources generate about 10 percent of the world's electricity.

Other types of renewable energy are geothermal, solar, biomass, and wind. The U.S. Energy Information Administration estimated that in 2015 renewable energy sources made up about 12 percent of all world energy used. That's estimated to increase to 17 percent by 2040.

CHAPTER THREE

ENERGY SOURCE ACCIDENTS

Energy sources, both renewable and nonrenewable, have benefited mankind. But they can also be destructive.

Nuclear energy can be extremely dangerous. The 1986 meltdown of the Soviet Union's Chernobyl nuclear power plant released large amounts of radioactive materials into the surrounding atmosphere and land. The disaster killed 28 people immediately, but thousands of others were, and are still, affected by the exposure to radiation. Because of this contamination, the Soviet Union created an exclusion zone surrounding the plant. This area may not be safe to live in for thousands of years.

Underground oil fields are under pressure, and specialized equipment is required to safely bring the oil up. In 2010, the Deepwater Horizon oil rig in the Gulf of Mexico exploded. This resulted in about 53,000 barrels of oil gushing up into the Gulf of Mexico every day for 87 days. The oil killed tens of thousands of fish, birds, marine mammals, and sea turtles.

There have been many disasters related to coal, mostly involving mining. The world's worst coal-mining disaster occurred April 26, 1942, in a coal mine at Benxi, China. Poor working conditions and a mixture of natural gas and coal dust caused an explosion that killed 1,549 miners.

Blackouts happen when electrical service to an area is disrupted. The largest U.S. blackout occurred after Hurricane Maria hit Puerto Rico on September 20, 2017. More than 100 days after the hurricane, as many as 675,000 people were still without electricity.

The offshore drilling rig Deepwater Horizon exploded in the Gulf of Mexico on April 20, 2010. This resulted in the largest marine oil spill in history. The explosion killed 11 workers and injured 17.

CHAPTER FOUR

THE GLOBAL ENERGY CRISIS

Some energy crises, or emergencies, occur because of increasing demands on limited fuel sources. Others occur because of a lack of other resources. In 2017, as many as 1.1 billion people around the world didn't have access to electricity. Reasons include poor infrastructure, poverty, and living in rural locations. Some people who do have electricity don't have a reliable supply of it.

There is progress, however. According to information from World Energy Outlook, in 2018 the number of people without electricity fell below 1 billion for the first time. Efforts to modernize old electrical systems and electrify rural areas are bringing power to more people worldwide.

Every year, more and more people are connected to electrical power. However, almost a billion people still don't have access to electricity.

As the world population grows, increasing energy demands on limited fossil fuel resources such as oil, coal, and natural gas are causing them to be used up more quickly. This could lead to an energy crisis.

In the 1950s, Marion Hubbert, a geologist working for the Shell Oil company, imagined that production of world oil would eventually hit what he called peak oil. After that point, production would begin to decline. Demand would begin to be greater than available supply. When this happens, the result will be higher oil prices. Because of other energy sources such as coal and natural gas, as well as new methods for removing oil such as fracking, peak oil hasn't been reached yet. However, it could be soon.

Since the mid-20th century, human activities have caused the planet to warm at an unprecedented, or unmatched, rate. Rising temperatures have resulted in serious consequences for the environment, upsetting the natural balance of ecosystems and pushing some species to the brink of extinction.

Before people understood the negative effects of burning coal and oil, they used these energy sources freely, with no controls or restrictions in place to prevent harmful gases from entering the air. These gases, smoke, and soot particles create pollution. Unchecked pollution in stagnant, or unmoving, air results in poor air quality called smog.

This picture shows President Nixon beside William Ruckelshaus, who was sworn in as the first administrator of the EPA in 1970.

In winter 1952 in London, England, foggy, stagnant air combined with emissions from coal fireplaces, automobiles, and factories. This resulted in the Great London Smog of 1952. The thick, stinking, brownish air covered the city for five days. The smog wasn't only unpleasant—it caused health problems in many people and led to at least 4,000 deaths. Over time, it may have caused as many as 12,000 deaths. As a result, England passed the Clean Air Act of 1956. Large cities started to switch away from coal as their primary energy source.

In the 1960s, U.S. smog events as well as coastal oil spills led to many protests and increased environmental awareness. As a result, the U.S. government took action. In 1963, Congress passed the Clean Air Act to regulate harmful emissions and decrease air pollution. Environmental efforts continued. In 1970, President Richard Nixon established the Environmental Protection Agency (EPA).

The United Nations, an intergovernmental organization dedicated to keeping peace worldwide, began to recognize the need to act internationally to address the issue of climate change. In 1997, countries made an international treaty, or agreement, called the Kyoto Protocol. Its goal was to curb harmful emissions, especially in larger, more developed countries where emissions are highest. The Kyoto Protocol targeted six main greenhouse gases, laying out regulations for the amount of emissions each country was allowed. Nearly 200 countries signed the protocol, but the United States dropped out of it in 2001. Former president George W. Bush argued that the Kyoto Protocol put unfair restrictions on the United States while leaving out developing countries.

During an international meeting held in Paris, France, in 2015, the United Nations Framework Convention on Climate Change agreed to increase efforts towards improving energy emissions and attempts to halt climate change. As of January 2019, 197 countries have signed the Paris Climate Agreement, which aims to keep the global temperature from rising more than 3.6°Fahrenheit (2°Celsius) from conditions before the Industrial Revolution.

DONALD TRUMP

On June 1, 2017, hundreds of people protested in New York City a few hours after President Donald Trump announced that the United States was withdrawing from the Paris Climate Agreement.

However, on June 1, 2017, President Donald Trump announced the United States was withdrawing from the Paris Climate Agreement. He said it put unfair restrictions on U.S. businesses. This is problematic, because the United States is one of the top producers of greenhouse gases. Research suggests that the time to curb the effects of greenhouse gases is limited to about 12 years. Rising global temperatures could lead to deadly consequences worldwide. A recent report suggested that the world should limit global warming to an even lower target of 2.7°F (1.5°C).

CHAPTER FIVE

CHANGING ENERGY SOLUTIONS

In the late 1700s and early 1800s, traditional modes of transportation, which typically required horses, changed to mechanical methods. Steam-powered cars appeared in the late 1700s. In 1855, Karl Benz in Mannheim, Germany, created the first gasoline-powered car manufactured for the public. American car manufacturer Henry Ford began selling his Model T automobile in 1908. By 1913, Ford revolutionized manufacturing when he began mass-producing Model Ts on assembly lines. With mass production, Ford's company was able to manufacture as many as 1,000 of the popular cars a day. As cars became more and more common, carbon emissions grew.

HENRY FORD

Assembly lines like this one at the Ford Motor Company plant in Highland Park, Michigan, revolutionized manufacturing and made automobiles more affordable.

For many years, the public was unaware of how carbon emissions contribute to global climate change. It wasn't until the 1970s that the growing environmental movement began to spread concern over and mindfulness about people's impact on Earth's climate. During the late 1980s, increased media coverage helped improve public awareness about global warming.

As people started to move away from using forms of fossil fuels for energy, electric-powered vehicles began gaining popularity. The first mass-produced hybrid electric vehicle was the Toyota Prius, released in Japan in 1997. A hybrid vehicle has a gasoline engine and an electric motor. In 2003, a group of engineers created the company Tesla Motors. Tesla now sells emission-free, battery-powered electric vehicles.

Despite growing awareness about climate change and progress in developing alternative energy sources for vehicles, there's still a long way to go. All forms of transportation combined to generate almost 25 percent of the world's greenhouse gas emissions in 2018.

Saving and storing electricity so it won't go to waste is an important step in lowering power costs and reducing harmful emissions.

After electricity is generated at a power plant, not all of it gets used. Some of the electricity goes to waste in transmission, or the process by which it's sent out. Some gets wasted in the distribution, or delivery, process. Overall, 6 percent of generated electricity was lost in the United States in 2013.

Electric companies also generate electricity based on demand, but it's not a perfect system. In order to always provide enough electricity, power plants generate more power than people actually need. The unused electricity sometimes goes to waste. There are several ways to store electricity, reducing this problem:

- Large batteries can store electricity to be used later.
- Machines pump water into reservoirs, or man-made lakes. When more electricity is needed, the water flows out, powering turbines and generators.
- Excess energy can be used to cool water or create ice. The cold water or ice is used to cool systems when demand increases.
- Kinetic energy can be stored by using large spinning devices called flywheels. Flywheel rotation can be used to power electric generators as needed.

Innovative energy storage solutions could help us limit our use of fossil fuels. Finding ways to conserve energy will be a key part of creating healthier energy practices in the future.

A flywheel electric power storage system is shown at the Tokyo Electric Power Station in Kofu, Japan. Global energy storage is predicted to increase by 12,200 percent by 2040.

People are increasingly using renewable energy types such as **wind**, solar, hydropower, geothermal, and biomass. Scientists are also coming up with advancements to these types of energy. New technologies are improving efficiency and lowering costs. This helps make it easier for people to adopt renewable energy in their day-to-day lives. Using wind, solar, and geothermal energy is more efficient and cheaper in 2019 than it was just a decade earlier. Hydropower has advanced less, but as of 2019, it's still the top renewable energy type used in the United States.

Scientists are also researching new ideas. People in Japan are exploring the idea of solar power from outer space. Since 50 to 60 percent of the sun's energy gets lost in the atmosphere, capturing the sun's rays in space would be much more efficient. In 2015, Japan's Aerospace Exploration Agency was able to send electric energy with microwaves. For space-based solar technology to work, we'd have to be able to transmit the captured energy wirelessly.

Space-based solar energy technology already captures the sun's rays and converts them into electricity on the International Space Station.

Scientists have learned to create electronic parts out of paper rather than plastic and other materials. Since paper products break down much more quickly than plastics, this could help curb harmful plastic waste and pollution.

Scientists are also experimenting with human power. Inventors from the United States and Canada have created a device people can wear on a knee. As the person walks, the device can generate enough electricity to power as many as 10 mobile phones. With over 7 billion people on Earth, that's a lot of potential energy.

CHAPTER SIX

ALTERNATIVE ENERGY MYTHS AND FACTS

Staying informed and separating myths from facts regarding alternative energy solutions helps people make better choices about energy use. Below are some common myths and facts about alternative energy sources and climate change.

Myth: Solar-generated electric energy is too expensive to save money or help the environment.

Fact: Older solar panels were only able to harness about 10 percent of the sun's energy. Newer solar panel technology and designs are lightweight, more efficient, and cheaper than older models. Some companies are manufacturing solar panels that look like traditional roof tiles.

Myth: Electric car batteries have short lifespans and can't go very far before they need to be recharged.

Fact: New lithium-ion batteries used in electric cars are very efficient and can store a great deal of energy. In 2019, electric car manufacturer Tesla announced that its Model S car can now travel 370 miles (595.5 km) on one charge. In November 2019, Tesla announced a new electric Cybertruck with a range of 250 miles (402.3 km) or more. Electric car manufacturers are also providing buyers with eight-year battery repairs or replacements.

Myth: Wind farms aren't efficient, are costly to build, and use more energy than they produce.

Fact: All sources of energy require money and materials to build. If wind energy was too costly, energy manufacturers wouldn't construct wind farms. A group of scientists studied wind power in 2010 and concluded that wind turbines create 20 to 25 times the amount of energy that's used to create them.

A Tesla Model S automobile recharges at a supercharger charging station. Supercharging stations are used to recharge a car traveling long distances.

CHAPTER SEVEN

YOUNG VOICES FOR THE ENVIRONMENT

Young people around the world are becoming involved in the fight to safeguard the future of the planet.

Kallan Benson of Crownsville, Maryland, first became interested in the climate crisis at the 2014 People's Climate March in New York City. She was only 9 years old at the time. She is now the national coordinator for the Fridays for Future movement in the United States. Fridays for Future is a student strike that was started by Swedish teen activist Greta Thunberg. Students take time off from school to demand action from their governments to curb global warming and climate change. Benson's environmental concerns and protests helped the state of Maryland pass a state-wide fracking ban.

Alexandria Villaseñor is a 14-year-old youth climate activist. She created an organization called Earth Uprising, which is committed to convincing world leaders to act now to keep the planet's warming below 3.6°F (2°C).

ALEXANDRIA VILLASEÑOR

On March 15, 2019, the Global Climate Strike took place around the world in 150 different countries. In Australia, 20,000 students protested at a rally, School Strike 4 Climate, where they demanded urgent action on climate change.

When 15-year-old Hannah Herbst of Florida learned her pen pal in Ethiopia didn't have electricity, she took action. Inspired by how ocean currents pushed her family's boat, she invented a device that captures ocean energy and converts it into electricity. Her small 3-D printed hydroelectric generator won the top prize at the 2015 Discovery Education 3M Young Scientist Challenge.

Whether by using your voice to speak up about climate change issues or conducting your own research to find new energy solutions, you can make a difference too. The choices you make today will have a lasting impact on the future of the planet.

CHAPTER EIGHT

PERSONAL CHANGES FOR CLEAN ENERGY

A person's carbon footprint is how much carbon dioxide or other carbon-based compounds they contribute to the atmosphere. In 2008, researchers at the Massachusetts Institute of Technology (MIT) came to several alarming conclusions. People living in the United States produce much more greenhouse gas emissions than the global average. A typical U.S. household produces 48 tons (43.5 mt) per year.

How you travel, what you buy, and even what you choose to eat can all contribute to your carbon footprint. Being mindful of how your actions affect the environment is the first step towards helping solve the current energy crisis. By committing to reducing your carbon footprint, you are conserving energy.

Switching from traditional light bulbs to more efficient LED lights is a good way to cut energy costs and reduce a family's carbon footprint at the same time.

Here are some suggestions for how to practice energy conservation in your home:

- Avoid buying or using single-use plastic items such as cups, bags, straws, and packaging material. Use reusable items instead. Reusable beverage containers, bags, and glass jars for food storage will help cut down on waste and reduce your carbon footprint at the same time.
- If you don't have far to travel, you can walk or bike instead of using a gasoline-powered vehicle. If you must drive, then carpooling or using public transportation are ways to reduce the number of vehicles on the roads.
- Buy locally produced food. When food is produced locally, it doesn't travel as far to get to you. Foods produced in other countries require extra energy to get to you. By buying local, you'll also be supporting your local community.

In November 2019, a government report stated that, "Earth's climate is now changing faster than at any point in the history of modern civilization, primarily as a result of human activities." Fortunately, new energy solutions can help reduce our reliance on the fossil fuels that are contributing to climate change.

Every day, people are exploring new energy innovations. Emerging technologies such as green hydrogen are being researched. Green hydrogen is a type of hydrogen produced by wind and solar energy. Since hydrogen production accounts for a large amount of CO_2 emissions, green hydrogen could be a useful alternative to traditional hydrogen, which is produced by natural gas and coal.

Some argue that investing in renewable energy solutions is too expensive or problematic. However, research indicates that clean alternatives are both cost efficient and productive. World leaders will need to work together to effectively curb the damage that fossil fuels have inflicted on the environment.

But beyond these new technologies and international efforts, there are important choices being made by everyday people. Practicing energy conservation in your home will help decrease the stress that's being put on the planet's limited resources. Simply unplugging your phone charger or turning off your computer can help.

You could even take your efforts further. Today, young people are at the forefront of the environmental movement. Teen climate activists are using their powerful voices to raise awareness about climate change and propose solutions to the current energy crisis. Your actions could make an impact in your community or even across the globe.

GLOSSARY

activist (AK-tih-vist) Someone who acts strongly in support of or against an issue.

contamination (kuhn-ta-muh-NAY-shuhn) The process of making something dangerous, dirty, or impure by adding something harmful to it.

efficient (ih-FIH-shuhnt) Done in the quickest, best way possible.

environment (ihn-VY-ruhn-muht) The natural world around us.

fossil fuel (FAH-suhl FYOOL) A fuel—such as coal, oil, or natural gas—that's formed in the earth from dead plants or animals.

generate (JEH-nuh-rayt) To produce something or cause something to be produced.

geothermal (jee-oh-THUHR-muhl) Related to the heat of Earth's interior.

greenhouse gas (GREEN-howz GASS) A gas in the atmosphere that traps energy from the sun.

hydroelectric (hy-droh-ih-LEK-trik) Related to the production of electricity by water power.

Industrial Revolution (in-DUH-stree-uhl reh-vuh-LOO-shuhn) An era of social and economic change marked by advances in technology and science.

infrastructure (IN-fruh-struhk-chuhr) The equipment and structures needed for a country, state, or region to function properly.

kinetic (kuh-NEH-tik) Related to the motion of something and the energy connected to it.

lithium-ion battery (LIH-thee-uhm–I-ahn BA-tuh-ree) A lightweight, high-power rechargeable battery commonly used in laptops, mobile devices, and electric cars.

radioactive (ray-dee-oh-AK-tihv) Giving off rays of light, heat, or energy.

stagnant (STAG-nuhnt) Not moving or flowing.

technology (tek-NAH-luh-jee) A method that uses science to solve problems and the tools used to solve those problems.

turbine (TUHR-byn) An engine with blades that are caused to spin by pressure from water, steam, or air.

INDEX

PRIMARY SOURCE LIST

Page 11
Eleven people missing after explosion at offshore drilling rig. Photograph. April 21, 2010. By U.S. Coast Guard.

Page 15
William Ruckelshaus being sworn in as the first administrator of the U.S. Environmental Protection Agency. Photograph. December 4, 1970.

Page 19
Assembly line at the Ford Motor Company's Highland Park plant. Photograph. Ca. 1913. Library of Congress.

WEBSITES

Due to the changing nature of Internet links, Rosen Publishing has developed an online list of websites related to the subject of this book. This site is updated regularly. Please use this link to access the list: www.powerkidslinks.com/SOGI/energy